Little Known Explorations

OVERLOOKED OR FORGOTTEN DISCOVERIES THAT NEVERTHE-
LESS HAVE FURNISHED FACTS THAT HAVE SERVED
TO UNLOCK THE SECRETS OF THE PAST

By HEREWARD CARRINGTON, Ph.D.

Author of "The Great Pyramid of Egypt," "The Seven Wonderr
of the World," etc.

HALDEMAN-JULIUS PUBLICATIONS
GIRARD, KANSAS

Printed in the United States of America

INTRODUCTORY

In the present book I have endeavored to cover some of the less-known explorations, referring only briefly to those which are more generally known—e.g., Polar explorations, discoveries in Egypt, etc. Aside from these, there are many others of extraordinary interest. Who, for example, first explored the interior of Greenland, Arabia, Persia, Tibet, New Guinea, the Libyan desert, Australia, Central Asia? Has the summit of Mount Everest ever been reached? Who first deciphered the wedge-shaped writing of the ancient Assyrians and Babylonians? These are some of the topics covered in the present book. I trust it may prove a useful addition to those already published in this series, dealing with

exploration.   H. C.

THE ROMANCE OF EXPLORATION

THE DECIPHERING OF THE ROSETTA STONE

Every now and then a dramatic discovery is made which catches the public imagination and receives wide publicity in the press. The unearthing of the tomb of Tutankhamen by Lord Carnarvon and Mr. Howard Carter is an example of this. It was spectacular and uncovered vast riches—hence perhaps the interest manifested. But there are scores of striking discoveries which have been overlooked or forgotten; and though all of them may not have yielded gold or silver ornaments, or baskets of jewels, they have nevertheless furnished us amazing information and supplied us with many Keys which have served to unlock the secrets of the past, and disclose the early history of mankind. The romance of exploration is a romance indeed—quite aside from buried treasure, which is occasionally discovered.

Probably everyone is more or less familiar with the story of the finding of the Rosetta Stone, and the part it played in the decipherment of Egyptian hieroglyphics. A French sapper was digging in the ruins of Fort St. Julian, near Rosetta, Egypt, when his pick struck a hard stone. Unearthing it, he found that is was covered with strange markings and, being a man of intelligence, turned it over to the authorities. It was a stone some 28½ inches wide, and 45 inches in length, and was divided into three sections, on each of which was writing. One of these was in Greek, one in demotic characters—the writing of the ordinary people in ancient Egypt—and one in hieroglyphics. No one in the world, at that time, had the slightest idea of what these hieroglyphics meant; not a single one of them had ever been interpreted. The "dead language" of Egypt was dead indeed.

Into the seemingly impossible task of translating these symbols, Dr. Young, of England, and later Francois Champollion, of France, threw themselves with enthusiasm. It was inferred (rightly as it turned out) that all three texts bore the same message. In studying the texts it was noted that two or three times certain groups of hieroglyphics were encircled, and it was guessed that these were proper names—probably the names of royalty. Fortunately these guesses proved to be right. So, letter by letter and sign by sign, the strange characters were interpreted. It took years of painstaking research to decipher the text in full; but when it had once been interpreted it afforded the key to the translation of other hieroglyphics, so that these could now be read, and even the famed "Book of the Dead" translated. It took more than a century of continuous work on the part of scholars to work out a complete dictionary of hieroglyphics; but this is now available, and the entire history of ancient Egypt may be studied with as much certainty and exactitude as the history of Greece or Rome. No more amazing task has ever been achieved by the mind of man. But, had it not been for the accidental discovery of the Rosetta Stone, we might even yet be completely ignorant of the meaning of ancient Egyptian symbols.

RAWLINSON INTERPRETS PERSIAN, BABYLONIAN AND MEDIAN CUNEIFORM

Equally remarkable—perhaps even more remarkable—was the interpretation of the ancient Persian cuneiform characters, which proved the key to the mysteries of ancient Babylon and Assyria. For here there was no Rosetta Stone to furnish an initial clue—nothing to help the decipherment of these strange inscriptions found on walls and tablets, which no man in the world could understand. They remained as great a mystery as ever, even years after the Egyptian hieroglyphics had been to a great extent interpreted.

The man who accomplished this seemingly impossible task was Henry Rawlinson. When but a lad of 17, he was sent to Bombay in the service of the East Indian Company, and eight years later he was placed in command of all troops in the Kermanshah district of Persia. The Orient had from the first cast a spell over him, and he mastered, with amazing ease and rapidity, a number of its languages and dialects. He was moreover a man of infinite patience and resourcefulness.

Behistun is about 20 miles from Kermanshah, and here, rising from the desert sands, is a huge rock, about 1,700 feet high. On the side of this, some 300 feet from the ground, had been carved a number of inscriptions. No man ever bothered to study these; they were completely inaccessible, for above and below them was the sheer cliff, which could only be scaled, at great risk, by ladders and ropes. At the foot of the inscriptions is a three-foot ledge, and below that the wall of solid rock, falling away to the desert below. Why should anyone scale this merely to look at the curious, wedge-shaped inscriptions on the rock above? And so for centuries caravans had passed it by, totally indifferent as to what they might mean.

Yet somehow Rawlinson felt that these inscriptions contained a vital secret, and he was determined to find out what that secret was. At great personal risk, he climbed to the ledge, at the foot of the writings, and on it placed a ladder, which enabled him to study the upper portions of the inscription. Here, perched high above the desert below, he copied and took impressions of the strange characters confronting him. A slip, or a sudden shift of the weight of his body, might have sent him crashing to his death. But, day after day and week after week, Rawlinson continued until his task was complete.

Then came the task of translation. With nothing to guide him, he undertook the work, and in 1846 published his great book, giving to the world for the first time the interpretation of the ancient Persian, Babylonian and Median cuneiform. The scientific world was incredulous; it declared that what he had accomplished was an utter impossibility. But it was wrong, and as the ancient tablets were unearthed—by the score, by the hundred—and their inscriptions read, it became evident that Rawlinson had indeed succeeded in deciphering these wedge-shaped markings. It was one of the most remarkable achievements in history.

The Behistun inscriptions are extraordinary in more ways than one. Over 500 B.C., Darius, King of Persia, had caused an account of his campaigns to be engraved on the rock in Persian, Babylonian and Median, so that all men who passed that way might read of the deeds of the great king. When finished, they had been coated with a varnish, some of which still remains, despite the passage of time, and the rain, hail and baking sunshine which had beaten upon it for 25 centuries. We have learned much, but the secret of that old Persian varnish is still a mystery.

LAYARD'S EXCAVATIONS IN NINEVEH AND BABYLON

Until 100 years ago, Babylon and Assyria were little more than names. Men had read of them in the Bible, but beyond that almost nothing was known of their civilization and culture. Nineveh and all their great cities had long since perished and been covered by the desert sands. None even know where they were situated. Credit for the first actual excavations there must be given to Austin Henry Layard, whose life was one long romance.

As a boy, he had been fascinated by the tales of the "Arabian Nights." He made up his mind that one day he would visit that country, live in it, explore it. Yet it all seemed such an utter impossibility. His father had decided that he must become a lawyer, and year after year, in the office he hated, he was compelled to carry on his studies, while dreaming of the lure of the Orient. Eventually, however, he passed his final examination, and was then more free to travel. Two events occurred in his life at precisely this juncture. He inherited some $30,000 from his mother, and an uncle of his returned from Ceylon, filled with enthusiasm for the marvels of India. Layard decided to travel thither. He sent one-half his fortune to a bank in Ceylon, and kept the other half for his traveling expenses. He decided to make the journey overland —for the Suez Canal had not yet been cut. So, in 1839, he set out, with a friend, to travel the intervening wastelands and deserts. In 1840 he arrived at Mosul.

Here he found that a Frenchman, M. Botta, was digging, but had discovered little of value. With his customary energy and enthusiasm Layard set to work. All sorts of obstructions were placed in his way by the Turkish government, and he finally succeeded in continuing his work only by bribing the soldiers who had been sent there to prevent him from doing so. Layard had a remarkable insight into the Oriental mind, and soon gained the friendship and even the admiration of the Arabs. His imperious personality won their respect, and on many occasions, when trouble arose, he rode directly into the camp of some insolent chief, and made him feel the weight of his words and even the strength of his arm. He constantly took his life in his hands—but his appeared to be a charmed life, and luck seemed to be with him in everything he undertook.

Digging in the great mound at Nimroud, he gradually unearthed countless treasures of bygone Assyria. First, clay tablets, vases and statues were discovered; then whole cities were laid bare. At Kouyunjik he found the palace of Sennacherib, buried 30 feet under an accumulation of debris and soil. He laid bare almost the whole of Nineveh. He excavated much of Babylon. True, these early explorations of his were carried forward and amplified by subsequent explorers, but the initial work was his, and had it not been for this it is doubtful if the later discoveries would have been made, at least for many years to come. Nebuchadnezzar's palace in Babylon might still be lying buried beneath the sands, and the uncovered Tower of Babel might still be considered a myth. All this and much more we owe to the initiative and determination of an obscure young law student—Henry Layard.

SCHLIEMANN DIGS UP CITY OF TROY

Amazing as were his discoveries, romantic as was his life, these in some ways pale, when compared with the discovery of the ruins of Troy, by Heinrich Schliemann, during the latter years of the last century. Schliemann's life might well form the basis for many a lecture on "applied psychology." His was a life filled with terrific adversities, culminating in one of the greatest triumphs of modern research.

Born in 1822, Schliemann was the son of a pastor of a small hamlet in Mecklenburg-Schwerin. All his early years were spent in abject poverty. For 18 hours a day he was forced to work in a small delicatessen store, waiting on the customers, sweeping the floor, cleaning the windows, and making himself generally useful. He was a true slave of the terrible conditions in those days. At the end of the day he was too exhausted to do anything but throw himself on a bed of straw and sleep until he was rudely awakened the next morning. Yet, while his body was thus occupied, his mind was recalling memories of the distant past. For, just as Layard had been fascinated by the "Arabian Nights," so Schliemann had been inspired by Homer's "Iliad." The legend of Troy entered into his very soul, and although the Trojan war was in those days considered entirely mythological, Schliemann believed implicitly in its actuality. "I shall find Troy," he said one day to a little girl who chanced to enter his shop. Years later he returned to find that little girl, only to discover that, with the passage of time, she had forgotten him and married another.

Schliemann craved knowledge, but he had no means of satisfying this craving. He could afford no books—had no time or energy to read them, even if he had been able to procure them. While selling food to his customers, his mind continued to be starved. Yet ever he dreamed on, thinking of Troy and his Trojan heroes.

One day a drunken miller lurched into the shop and suddenly began to recite, in Greek, some passages from Homer. He too was the son of a clergyman—condemned to be a miller. The meaning of the words was lost to the boy, but the mere beauty of the lines thrilled him. "Say it again," he pleaded, and once more, "say it again." Digging into his pocket, he extracted the few pennies he had in the world, in order to buy a drink for the drunken miller, so that he might hear the lines once more. Tears rolled down his face, and all that night he lay awake, tossing on his bed of straw, thinking of the heroes of the Trojan war.

Then a tragedy occurred. Schliemann injured himself while lifting a heavy cask and, being unable to work, was promptly discharged. Not knowing what to do he drifted to Hamburg. There he shipped as a cabin-boy on a vessel bound for Venezuela. The ship was wrecked in a storm, and for hours the crew of the boat faced death, before being cast up on the Dutch coast. Here for weeks he was reduced to absolute beggary. He finally secured a poorly paid job in an office, which served to keep him alive. He rented a garret for 30c a week, and starved himself in order to buy books for study. Twenty-five cents a day sufficed to feed him and pay his rent. The rest he spent on books.

Now at last his mind began to receive the "food" it had long craved. He read constantly; he was never without a book in his hand. In six months he had taught himself English; in the next six months he had mastered French. His genius for languages was phenomenal. He taught himself Spanish, Italian, Dutch, Portuguese; he even took up Russian, and in six weeks was able to write a letter in that language. This seemed almost fore-ordained, for a few weeks later he was sent to Russia on business—and decided to settle there and make his fortune. His sole object in doing so was to obtain the necessary freedom and money which would enable him to excavate the ruins of Troy. Within 10 years he was a wealthy man—and meantime he had mastered Greek.

Now, at 35, Schliemann set forth on his quest. The discoveries of Layard and others had fired his imagination, and he determined to devote his life and his fortune to the discovery of Troy. His studies had suggested to him its possible situation, so in 1870 he bought the greater part of the desolate Hill of Hissarlik from the Turkish government, and there began his digging the following year. Scoffs and ridicule had no effect on him; he expended his energy and his money freely. The world still believed that the legend of Troy had been invented by Homer, and that it had no real existence. Even those archeologists who took it more seriously still insisted that, wherever the ruins of Troy might be, they were certainly not buried beneath the Hill of Hissarlik. Disregarding all such criticisms, Schliemann continued his gigantic task of excavation, moving millions of tons of earth and carting them away to considerable distances. Truly, his was "the faith that moves mountains."

An then, one day, the spade of one of his diggers struck a stone—a stone wall. Feverishly now the digging was continued. Deeper and deeper his workers bored into the hill, and the deeper they dug the more marvels of the past they disclosed. Here, surely, was the most marvelous hill in the world. City rested upon city, civilization upon civilization. Under the first city lay a second, and under the second a third, until in all *seven* cities were discovered, one resting upon the other, with debris and ashes between them, often many feet in thickness, telling of the disasters which had overwhelmed them. Here were the houses, the walls, the ruined temples; here were treasures of all kinds—golden cups and silver goblets, necklaces and jewels, all the relics of bygone civilizations. Here at last was his city of Troy, which as a boy he had vowed to unearth.

For three years Schliemann dug into the Hill of Hassarlik. He discovered walls eight and 10 feet thick; he laid bare an ancient gate; he found a city of Troy which had flourished 3,000 years before the Greeks took it by subterfuge—buried under 30 feet of debris. Later he unearthed, at Mycenae, an even more wonderful treasure than that of Troy—the bodies of ancient Kings buried in golden masks and with golden armor. Here indeed was the wealth of the fabled Mycenaean Age, which had similarly been considered mythical by the skeptical world. When Schliemann published his discoveries, they were met with a storm of criticism and at first no one believed him. The same old story! But time and further research showed that Schliemann was right and that his critics were wrong. He had indeed discovered the ruins of ancient Troy, and he lived to see his enemies confounded. Working in direct opposition to the opinions of others, he had dug up the city in the very place where he felt it must be, and where most scientists had said it could not possibly have been. The dreams of the poor grocer-boy had found their fruition in one of the most romantic and spectacular discoveries of our age—or any other age.

THE RECOVERY OF THE "ELGIN MARBLES"

About 150 years ago, and about the time that the Rosetta Stone was brought to light at Fort St. Julian, Lord Elgin, the British ambassador to Constantinople, made up his mind to rescue all that was possible of the relics scattered about the Acropolis at Athens (then ruled by the Turks) and whatever other art treasures he could find. Many of these had been wantonly destroyed by the Turks, and there was every evidence that the remainder would share their fate ere long. He was a wealthy man in his own right, and was willing to undertake the work of excavation if he could obtain permission to do so. He accordingly applied to the Porte, and after much deliberation he was finally granted permission to explore the Acropolis and surrounding territory upon payment of $25 a day. This he willingly paid out of his own pocket, and the work was begun.

A rumor reached his ears that some of the famous marbles had been built into a certain Turkish house. He finally obtained permission to pull this down, after paying and bribing its owner to let him do so. When it was completely demolished, no trace of the marbles had been found, and it was then that the wily old Turk calmly informed him that he had ground down the marbles to make mortar for his dwelling. This was but a sample of the many difficulties and disappointments which Lord Elgin encountered during the course of his undertaking.

Despite these, a number of the marbles were collected and shipped back to England on a boat named the "Mentor." The vessel sank, however, off the Grecian archipelago in some 60 feet of water. It took three years to salvage the marbles, with the aid of divers. Finally this was done and they were placed aboard another boat. This reached England in safety, but the problem then arose as to what should be done with them. Lord Elgin had spent a large part of his fortune in these excavations, and he wished to sell them to the British Government for the

Museum. What were they worth? Who could say? Famous artists were consulted, but declined setting a figure on objects of art which were priceless. Finally they were acquired for the sum of $175,000, which was but a fraction of their real worth. Today they are in the British Museum, where they have been seen and admired by countless thousands of visitors. The world owes much to Lord Elgin for having rescued these beautiful relics of Greek art, before they were utterly destroyed and lost to the world forever. Had he lived a century after he did, it would have been too late.

EGYPTIAN TOMBS AND MUMMIES

The Valley of the Tombs of the Kings is well known to every tourist, and the discovery of Tutankhamen's tomb created a flutter of interest throughout the entire world; but it is not sufficiently realized that, long before this, burial places and mummies had been constantly unearthed. As far back as 1881, Maspero had made a veritable "haul" of mummies—11 kings, nine queens, a prince and a princess, all in one day. This was near Thebes. Here were Seti I, Thotmes I, Thotmes II, Thotmes III, Rameses II. Vast treasure was found in this burial place. It was one of the most dramatic discoveries in all history.

The infinite pains taken by the ancient Egyptian rulers to hide their remains and prevent their tombs from being plundered is well known. The utmost secrecy was employed, all sorts of obstructions were placed in the way of robbers, and the secret entrances were so skilfully hidden that they often remained undiscovered for centuries. Yet all their labor, all their secrecy was in vain. Hardly a tomb in all Egypt has been found intact. Almost every tomb discovered has been rifled of its contents. The terrible curses which were launched did not prevent their desecration, for the thieves were evidently quite prepard to sacrifice their chances in the next life for the prospect of gaining something in this. It is fortunate, however, that these thieves were more interested in jewels than in mummies, and many of these have now found their way into the museums of the world.

One of the most dramatic—and little known—discoveries ever made in Egypt was Professor Flinders Petrie's finding of the mummy of Horuta at Hawara, at the bottom of a well 40 feet deep. Here in a flooded chamber, amid impenetrable darkness, he and his laborers wrestled continually with mighty blocks, in order to get to the stone sarcophagus which he suspected was there. They found it at last, with the lid barely peeping above the surface of the icy water.

For days they strove to shift it, but it was immovable, so he decided to cut the sarcophagus in halves in order to get at the inner coffin. Weeks of fatiguing labor saw this gigantic task accomplished, and there was another desperate fight, with men working up to their chests in water, to get it out.

The foot-end came to light first; but the coffin remained in the other half, and was apparently as far off as ever. The Egyptologist, groping in the murky water, fought with it, strove to shift it with his hands, with his feet. It was firmly fixed.

Still he was not beaten. After a sustained effort lasting several days, he and his workers managed to raise the lid of the other half of the sarcophagus with wedges, until the inside of it was a few inches above the water level. Then he wriggled inside, and for hours in the darkness he sat astride the coffin and struggled to loosen it. The top of his head touched the lid of the sarcophagus, and he had hardly room to move at all; the water came up to his mouth and compelled him to breathe through his nose. More than once in the course of his tremendous exertions he took in a mouthful of the nauseous water. The sand clung to

10

the coffin as though it were set in a bed of cement. He tried scraping away the sand with his feet, he pried at the coffin with crowbars. All his efforts failed to shift it a fraction of an inch.

Few men would have continued under such hopeless conditions; but Flinders Petrie was possessed of a determination that would not be denied. He set to work drilling holes in the coffin—a most difficult feat. When this was done bolts were inserted, strong ropes were attached, and the men went along the passage and hauled away with all their strength. For a time it was like heaving at a mountain; then the coffin stirred slightly, moved more and more. Backs were bending under the strain, arms almost cracking as the men taking part in that fantastic tug-of-war with a dead man finally triumphed, and dragged the water-blackened coffin out of the depths.

Breathlessly they opened it, and found the mummy of Horuta—"wrapped in a network of lapis lazuli, beryl, and silver . . . Bit by bit the layers of pitch and cloth were loosened, and row after row of magnificent amulets were disclosed, just as they were laid on in the distant past. The gold ring on his finger, which bore his name and titles, the exquisitely inlaid gold birds, the chased gold figures, the lazuli statuettes, the polished lazuli and beryl and carnelian amulets finely engraved."

The discovery of the mummy of Horuta is one of the epics of the Nile.

To deal further with the romance of Egyptian exploration would however take us too far afield, in a general little book such as this. The reader will find it more fully outlined in Little Blue Book No. 469, to which he is referred, as well as to many of the writings of Mr. Joseph McCabe.

EXPLORING THE LIBYAN DESERT

Many legends are mixed with the sands of the Libyan Desert, one of the most fascinating of these being the lost army of Cambyses, who sent it against the oasis of Jupiter Ammon, only to be swallowed up in the vast wilderness and disappear forever. Then there is the legend of Zerzuza, or the Little Birds—a mysterious white city, filled with innumerable riches, where slumber a king and queen who must not be disturbed, and where there are many beautiful, ancient drawings and works of art. Such are the tales handed down by the Arabs.

Naturally, all such stories were regarded as mythical. However, during World War I, an advance patrol sighted in the interior of the desert a huge plateau, 3,000 feet high, with steep, rocky sides, precipitous and dangerous. This was named the Gilf Kebir and, although it can hardly be seen on a map of Africa, is nevertheless about the size of Switzerland. It lies south of the great sand-sea, and is about half way between this and Kufara, reached by Rosita Forbes.

A Hungarian count, Ladislaus de Almasy, had become greatly interested in the riddle of Zerzura, and this had been stimulated by the discovery of ancient rock-drawings by both Hassanein and Patrick Clayton, who had independently explored the interior of the desert. Accordingly, in 1932, an expedition was organized by Almasy, in which Clayton and Squadron-leader Penderel joined. They found that they were unable to climb the Gilf, but, in a flight over it, palm trees had been seen, which discovery led to the organization of another expedition, the following year, better equipped for the undertaking.

About half way to the Gilf they came upon a place known to the Arabs as Abu Ballas—the Father of Jars. Sure enough, at the foot of a hill, were a number of jars, some of which showing the unmistakable shape of the classic Greek *amphora*. Evidently this spot had been selected as a sort of half-way camp in ancient times, where water was stored for passing caravans. Herodotus tells us that this had been the practice of the Persians during their campaigns. Had these jars any connection with the lost army of Cambyses?

11

From an old caravan-guide, it was learned that there were several *wadis;* one which had already been glimpsed, a second, known as the **Wadi of Acacias**, and a third, known as the **Red Wadi**. A circumnavigation of the plateau was then begun in the cars, in an attempt to find some slope which would permit their ascent. Finally one was discovered, and the explorers reached the top. Here they made an interesting discovery—a circle of stone pillars, showing that the place had been well populated in times past. In a near-by cave they found a number of stone-age implements, and on the walls drawings of lions, giraffes and other animals, in four colors. At the more southern oasis they found dozens of caves, all filled with rock-drawings—a veritable prehistoric art gallery. Among these drawings was one of a swimmer—showing that a lake had been there at one time; and indeed there were many indications of former fertility. There is therefore evidence that the Gilf was once a thickly populated and thriving area; and it seems more than probable that these facts may have given rise to the Arab legends—of ancient cities and green oases in the Libyan Desert.

LOST CENTRAL AND SOUTH AMERICAN CIVILIZATION

Although it has been generally known that the Mayas, the Toltecs, the Aztecs and the Incans possessed a magnificent culture and civilization, and although it was known that they constructed extraordinary buildings, it was not realized until the past few decades how truly remarkable these were—the beauty and massiveness of the streets, causeways, temples and walls which they built. These unfortunately are nearly all that remain to tell us of the glories of the past, for the conquering Dons stole and ravaged and despoiled everything they could lay their hands on; only the ruins of these structures remain, for the simple reason that they were found too tough and enduring to be destroyed.

Volumes could be written on these ancient civilizations, but one or two examples must suffice. The Incan Temple of the Sun at Cuzco was doubtless one of the most beautiful structures in the world. It was built of immense blocks of stone, fitted so perfectly that even today, after a lapse of countless centuries, it is impossible to insert a six-thousandth of an inch gauge between the stones. No two blocks were alike in shape and size, but each was most accurately fitted to those about it. No modern mechanic could produce results more minutely accurate. Yet, though they weighed many tons, they were somehow hoisted up to the top of the great walls and there emplaced.

When the Spaniards found it, the temple walls were covered, inside and out, with burnished gold. An immense, gem-studded Sun of gold stood at one end, and opposite this was a gigantic moon of polished silver. Beneath the image of the Sun were the mummies of the 12 Incan emperors, covered with gold and glistening with jewels. Every available space in the temple was filled with priceless statues and objects of art.

The Spanish conquerors, "with sword in one hand and crucifix in the other," ruthlessly tore these relics from their places; the regal mummies were thrown down, hacked to pieces and their regalia and ornaments torn off. Tapestries were ripped to bits and destroyed. Loot was carried from the temple by the carload. Of that vast treasure of the Temple of the Sun, all that remains intact today are a few bent and battered plates of thin gold that once formed part of the covering of the outer wall, and which were dropped, trodden into the earth and overlooked by Pizarro's men.

Throughout all central and south America it was the same story. Everything that was capable of being removed was carried away, and the rest was destroyed, if that was possible. Is it any wonder that but few relics of these bygone glories remain to us? Only ruins stand as mute evidence of the wholesale destruction that was carried out. But even these are sufficient to tell us that millions of men and women must have lived in these fertile regions in ages past, and that their culture was one of the most extraordinary that ever existed upon the face of our globe.

Further archeological investigations will doubtless add much to our knowledge of these civilizations. The reader is referred to the writings of Mr. Joseph McCabe (published by Haldeman-Julius) for a most interesting and enlightening summary of such discoveries to date.

EXPLORING AMERICA'S NORTHERN COAST

THE WORK OF EIELSON AND WILKINS

It is not generally realized that the whole of the northern-most coast line of Canada, as well as that of Alaska, lies within the Arctic Circle. Much of this country is relatively unexplored, and the winters are as cold as northern Greenland. Exploration of this territory, and of the waters lying directly north of it, is therefore almost as hazardous as Polar exploration, and many are the ships and lives which have been lost in attempts to chart this region and to discover a northwest passage.

Several of these later explorations have been made by air. George Wilkins flew several times from Fairbanks, Alaska, to Point Barrow, on the north coast, and out over the Arctic Ocean. On the first trip Barrow was missed in the fog and Wilkins and his pilot Carl Eielson flew well out to sea before discovering their mistake. On the return journey they passed a village which was not marked on the maps and, being unable to land there, they dropped a note, asking the villagers to outline the name of the village in the snow. They did so, but still the name was unknown to them. They accordingly dropped another note, asking that they make an arrow, pointing in the direction of Fairbanks. The villagers formed themselves in an arrow, their dark bodies being clearly distinguishable in the snow. Guided by this the explorers continued their journey, finally arriving safely at their destination.

On his third flight, Wilkins had one of those narrow escapes from death which have so often occurred. Peering through the mist, he suddenly saw to his left a high rocky mountain peak, and rapidly signalled his pilot to turn to the right. The latter signalled back that there was one on the right also. The machine was so heavily loaded that they could not hope to rise above these peaks, and their only hope seemed to be to fly between them, trusting to luck that they could clear the intervening mountains. As they flew on, however, the pass narrowed, until there were only two or three feet on either side of the wings. On the topmost ridge the wheels actually touched the snow and started to spin, just as they would do after a take-off. However they were through, and eventually landed safely.

The fourth flight, made the following year, nearly proved fatal. Engine trouble developed and they were forced to land on the ice. Both men worked frantically on the engine for two hours, in the icy wind, before rectifying the trouble. Ten minutes after the take-off, the engine was sputtering as badly as before, and a second landing was necessitated. Eielson worked steadily in the 30 degrees below zero temperature, with four of his fingers frozen solid; they had to be amputated later at Barrow. After more than an hour's work, however, the engine again ran smoothly; but it had begun to snow, and this proved so sticky that the plane refused to rise. By following their former tracks, however, they finally took-off, on the very edge of a bad patch of ridges.

All these take-off attempts had consumed much gas, and it now seemed doubtful that they would have enough to reach Barrow. The

13

only thing to do was to keep going, hoping that the plane would carry them over the most dangerous terrain. Suddenly the engine stopped dead. Eielson steadied the machine into a glide, and a moment later it crashed into a pile of snow, which bent the skis and twisted the stanchions. All hope of flying back to their base had to be abandoned.

It was now dusk and the navigators did the best they could to make themselves comfortable for the night—after sending out a radio message (which was never received). The following morning they discovered that the ice on which they had landed was drifting away from Barrow at the rate of five or six miles an hour. They therefore improvised some sleds, from parts of the plane, loaded them with what provisions they had, and started off on foot for a trading-post, about 100 miles away.

For four days they were unable to move because of a blizzard, but on the fifth they set out, making slow progress over the pack-ice. The sleds were abandoned and the loads carried on their backs. Once Wilkins fell through the ice into the water. He pulled himself out by means of his ice-pick and his companion pulled off his trousers and boots, which had been frozen solid. Fortunately they had some dry clothes with them, so, donning these, they again set off, walking as fast as possible to prevent his feet from freezing. For 18 days they continued their march doggedly, Eielson's fingers all the time causing him more and more agony. Finally they sighted the dim outlines of the settlement, and three hours later were eating their first hot meal in nearly three weeks. It had been an uncomfortably close call.

Perhaps one might think that our explorers would have had enough of the Arctic by this time? Not a bit of it! The next year Wilkins and Eielson set sail again—in a new plane. This time they were forced down by a terrific hurricane. A rapid landing on the ice had to be made, the tanks quickly emptied of gas, so that they would not freeze, and the plane banked-up with snow, so that it would not turn turtle in the wind. Once again, for five days, they were confined to the cabin by the storm, but on the sixth day the weather cleared and they were able to take some observations. They found that they had landed on the west coast of Spitzbergen. Carefully collecting every drop of gas, they prepared a 100-foot runway—but the machine would not move. Wilkins climbed out and gave the tail a push. The plane started and Wilkins attempted to climb aboard by means of a rope ladder, which had been provided for the purpose. He clambered onto the tail and tried desperately to reach the cockpit. The plane was gaining momentum, and Wilkins threw off his mittens, in order to get a better grip on the rope. But his hands were so numbed that he could not hold on, and clung to the rope with his teeth. Finally, realizing that he could not make the cock-pit, he flung himself from the plane onto a snow bank, just before the plane left the ground. The tail struck him and left him half stunned.

Eielson, seeing this, circled back and once more landed on the ice. By this time Wilkins had collected his faculties, and a third attempt at a take-off was made. This time Wilkins hooked one leg in the cock-pit opening, and pushed with a log of driftwood which he had found. With a sudden lurch the plane started, and the explorer tumbled into the cabin. It was fortunate indeed that no more landings had to be made, as the fuel was running dangerously low. But they did not have far to fly before sighting Green Harbor, and a little later they were safe once again in the potposts of civilization. They had flown from Alaska to Spitzbergen for the first time.

AMUNDSEN AND THE MAGNETIC NORTH POLE

Amundsen is another name connected with the exploration of these waters. At the beginning of the present century he had conceived the plan of discovering the magnetic north pole, and endeavored to raise

14

the necessary money for such an expedition. It took him three years to do so, however, and he finally set sail, with a crew of six, on a 47-ton fishing-smack, named the *Gjoa*. They reached northern Greenland safely, but here their troubles began. The small vessel was grounded on the rocks, during a heavy storm, and it was found impossible to dislodge it. It had been greatly overloaded with supplies, and, after the hold had been filled, the whole upper deck was piled high with boxes of food and other supplies. The only course seemed to be to throw much of the precious cargo overboard, and this was accordingly done. Finally a heavy wave freed the vessel, and she floated back into deep water.

Then a fire started in the engine room, perilously near 2,000 gallons of gasoline, which had been stored in cans. The men stuck to their posts, manned the pumps, and the fire was eventually extinguished. Next, the ship ran onto a large submerged reef, and to free it, more cargo had to be sacrificed. Even then their troubles were not over. For the next five days it blew a terrific gale, and there was danger that the anchors would drag and the ship be dashed ashore. At the end of that time, however, the wind abated and they were able to reach the harbor safely, which was destined to be their home for a summer and two winters.

Throughout their explorations the Eskimos proved most helpful and friendly. The westward voyage was eventually begun, but the vessel was in constant danger, owing to submerged rocks and reefs. Sometimes a small boat had to be sent ahead, to take soundings, before the vessel could proceed. Finally an American whaling ship was sighted, and it was evident that the passage of the northwest channel had been successfully accomplished. In August, 1906, they dropped anchor at Nome, after having been forced to complete the last leg of the journey by her sails alone—the propellor having been smashed by a hidden rock some time before.

Stefansson is another name associated with explorations in these parts of the world. He learned to live like the Eskimos, and in his book, "The Friendly Arctic," he endeavored to dispel the illusion that lands above the Arctic Circle are always dark and ice-covered—stating that plant, bird and animal life are abundant there, and that the temperatures in the heart of Montana are often colder than they are at the North Pole. In the summer, they sometimes rise to 90 degrees in the shade. He pointed out that the only way to live in the Arctic is to live like the Eskimos.

Stefansson set sail from Nome in three ships, the Karluk, the Mary Sachs, and the Alaska. He was in personal command of the first of these, the largest of the three. Unfortunately this ship became trapped in the sea-ice, and nothing they could do would free it. Stefansson decided to go inland on a hunting trip, and set out, accompanied by four white men and two Eskimos. The next day, however, a fierce gale sprang up, and Stefansson was astonished to see the Karluk begin to move eastwards. Then mist clouds and snow concealed it, and when these cleared away the vessel had disappeared.

The party set out along the coast on sleds, in search of this ship, and the others which had not been sighted. He finally came upon the two latter vessels at Collinson Point, and from there he carried on a number of explorations, discovering several islands which were hitherto uncharted. It was during one of these journeys that he encountered some of the crew of a whaling vessel, the Polar Bear, and some casual reference was made by them to the war. "What war?" asked Stefansson. He was astounded to learn that the greatest war in all history had been raging in Europe for more than a year.

News finally reached him of the Karluk—and it was bad news. The vessel had been trapped in the ice and had sunk; most of the crew had perished, though a few had reached land in safety. Stefansson himself continued his explorations, and, after more than five years in the Arctic,

15

finally returned to civilization, having discovered several new lands and charted thousands of square miles north of Melville Sound, off the north coast of Canada. The northwest passage and the adjoining waters had been explored for the first time, and were added to the maritime charts of the world.

POLAR EXPLORATIONS

A Little Blue Book has already been devoted to this subject (580), so I shall not deal with it in the present book, beyond a few brief references, for the sake of inclusiveness. Leaving aside the dubious claims of Captain Cook, the first authentic date on which the North Pole was reached was April 7, 1909, by Peary. Numerous observations were made and photographs taken, and it seems certain that the spot whereon the flag was planted was indeed the Pole, which many brave men had tried vainly to reach for more than 100 years.

The Amundsen-Ellsworth expedition in 1924 came within 90 miles of the Pole, but on May 12, 1926, Amundsen flew over the top of the world in his airplane, the Norge. Meanwhile, on May 9, Byrd had flown over it, being the first to do so. Since then, other flights have been made.

The South Pole had always been more inaccessible than the North, partly because of its great distance from any habitable mainland. Yet attempt after attempt had been made to reach it. Older explorations aside, Commander R. F. Scott made the first attempt in 1901, but was forced to turn back on reaching the great ice barrier. On January 16, 1909, the south Magnetic Pole was located by Mawson and David. Meanwhile Shackleton and Mawson had organized and carried out new expeditions. A famous French explorer, Dr. Jean Charcot, had navigated his ship the "Pourquoi Pas?" to the most southerly point ever reached thus far by a vessel. But none of these expeditions reached the South Pole.

It was eventually reached for the first time on December 14, 1911, by Amundsen. On January 16, 1912, Captain Scott's ill-fated party also reached it—just one month after Amundsen had planted the Norwegian flag there. The tragic and heroic end of that expedition is well known. Since then, Admiral Byrd's historic explorations were all carried to a successful conclusion, and he lived for several months at the Pole, making daily observations. An interesting account of his sojourn there may be found in his book, "Alone."

SOME LITTLE KNOWN EXPLORATIONS

BURKE AND WILLS EXPLORE AUSTRALIAN DESERT IN 1860

We who live amongst the comforts of civilization are inclined to think that only a few stray corners of the world remain unexplored. As a matter of fact, until our own generation millions of square miles of the earth's surface had never been seen by white men, and even today there remain vast areas which are totally unknown. The present century has contributed much to our knowledge of these blank spots, and polar exploration has received wide publicity in the papers—so that such names as Byrd, Peary, Amundsen, Scott, Stefansson and others have become almost household words. But what of the explorations of other lands? Little seems to be known regarding them. It is of interest to know who first explored Australia, Greenland, New Guinea, Persia and many other countries; whether the summit of Mount Everest has ever been reached, and if so by whom, and so on. Relatively little seems to be known of these men and their exploits, which in many cases involved immense difficulties and hardships, rivaling in fantasy the stories of the "Arabian Nights." But in these cases their adventures were real.

Let us begin with a simple, straightforward case. Civilization clings almost exclusively to the coast-line around Australia, and the millions of square miles of its interior are still but little known. In 1860, Burke and Wills first crossed its great deserts, reporting an almost total absence of surface water or oases, which make them even more difficult of passage than the Sahara. A few "soaks" or water holes were found, and it was evident that water must exist somewhere, since the bush blacks manage to live there. But it is a forbidding country. Later, Mackay and Weston explored parts of the interior. The most famous of these Australian explorers, however, is doubtless Michael Terry who in 11 years (1923-34) undertook no fewer than 12 expeditions, during which he made many excellent maps.

The first crossing was made in an old Ford car, and on several occasions they lost their way, and experienced hunger, thirst and forest fires. The natives in many parts were also definitely hostile, and several of the outlying settlers had been killed. Subsequent journeys were, however, far less difficult, when well financed and adequate expeditions were undertaken. Due credit must be given to Michael Terry for his Australian explorations, but they do not include the dangerous, hair-raising adventures which some other explorers encountered.

KOCH AND ERICHSEN IN THE ARCTIC CIRCLE

Take for example the experiences of the explorers of Greenland. In 1906, Captain Koch and Mylius Erichsen set out on an expedition to map the interior, setting up a base camp at Danmark's Harbor, some 700 miles above the Arctic Circle. After some preliminary surveys, Erichsen decided to explore to the westward, while Captain Koch returned to the base camp for supplies. Erichsen's party perished to the last man, some of their bodies being discovered by relief parties in the spring of 1908. Erichsen and some of the others were not found, however, and in 1909, Captain Einar Mikkelsen set sail from Denmark, on the "Alabama," in an attempt to discover them. They reached Danmark's Harbor and carefully explored the surrounding territory—without however finding any further traces of Erichsen's party. They spent the winter there and, the following spring, began the ascent of the steep slope of the ice-cap, with 15 dogs and provisions for 100 days. They discovered camp sites and messages left by Erichsen, but no bodies. Towards the end of May, Mikkelsen decided to return to the ship, as the sun was melting the snow, and a number of accidents had reduced them to seven dogs and one sled. Moreover Mikkelsen was suffering from scurvy, and could scarcely move one leg before the other.

By midsummer pools of water were forming in all directions, and the food was almost gone. For days they had to depend on sea gulls for food. Huge cracks were forming in the ice, and they had to construct a raft capable of floating the sled. The dogs began to die off, and the explorers were forced to eat their livers—knowing that the meat would have a toxic effect. The water was now so deep that they had to abandon the sled. They still had 60 miles to go to reach Danmark's Harbor, with only three pounds of food left. It had become a race against death.

Reaching a fjord, they found it open water and had to make a long detour. Then a blinding snow storm came on, and for 40 hours they had to seek what shelter they could among the rocks. Several of the men became delirious and experienced strange hallucinations. Finally the gaunt survivors straggled into Danmark's Harbor, where they found a depot of food.

Winter however was now fast approaching, and they still had 100 miles to go to reach the "Alabama" at her winter quarters near Shannon Island. Struggling forward again, through foul weather, they at length reached their promised haven—only to encounter their greatest disappointment of all. Instead of the vessel, they found only a hut built from its timbers. The ship had been wrecked and the crew had been taken home on another vessel.

There was nothing for them to do but pass the long winter in the hut. When spring came they made their way north again to recover their lost belongings. All that summer the castaways waited for some relief; but with the coming of the autumn they realized that they must resign themselves to another winter in the hut.

The ensuing months were one continuous battle against cold and hunger. Their hands and feet were frozen, and they had no means of heating the hut. To add to their dangers, foxes and polar bears constantly threatened to attack them. On one occasion a bear actually smashed in the door; their rifles were frozen and their only weapon was an axe. Fortunately, however, one of the rifles still worked, and a well-directed shot brought him down before he could kill the occupants.

The sun returned in February and the explorers made their way to Bass Rock, which they considered the most likely port of call. Here they waited week after week, keeping a constant lookout for a friendly sail. At last, one morning, they awakened to see a small steamer lying off the rock. Their calls and frantic signals were finally noticed, and they were taken aboard, and finally reached Denmark, after an ordeal such as few travelers in the Arctic have survived.

HENRY GEORGE WATKINS' FATAL VISIT TO GREENLAND

Henry George (Gino) Watkins was another famed explorer of Greenland. He too underwent a series of almost unbelievable hardships and privations, but was less fortunate than his predecessors, for he lost his life while exploring the foot of a glacier in a small boat. His companions saw the boat floating low in the water and searched for him until midnight, and all the next day, to no avail. An accident of some sort had evidently happened, and Gino Watkins found his grave in the icy waters of the Arctic.

How amazingly different are the adventures of those intrepid explorers who have undertaken expeditions into the vast jungle lands of the tropics. Their hazards and hardships have been equally great, but how different in character. Instead of the icy cold and blinding snow, they encountered suffocating heat, hostile natives, torrential rains and the bites of insect pests of every conceivable variety. Truly humanity is tough, and how many of these explorers survived at all is little less than a miracle.

EXPEDITIONS INTO DUTCH NEW GUINEA

Take, for instance, the expedition into the interior of Dutch New Guinea, undertaken by Captain Rawling, in 1910, assisted by the Dutch government. This is one of the largest islands in the world, being some 1,500 miles long and 400 miles wide. The northern and southern portions of Dutch New Guinea are divided by a high range of mountains, culminating in a peak known as Carstensz Top, about 1,600 feet high. It is snow-covered, but the jungle lands leading to it are malaria-ridden and full of natural obstacles. For these reasons it had remained largely unexplored until the Rawling expedition was undertaken.

As porters and carriers the Dutch government had provided 40 Javanese soldiers and about 60 convicts—many of them arriving in chains. Some of them were convicted murderers. The vessel which bore the party along the coast and up the Mimika River was the Nias. It was soon found, however, that river navigation was impossible, and so a base camp was established and the journey across country begun.

This soon brought them to the Kapare River, but at this point a number of the Papuans deserted, being afraid to go further. To shame them, Rawling shouldered a pack and started off up the river bed alone. There was a sudden gutteral cry, and several of the natives disappeared into the jungle. Rawling continued, thinking no more about it. But a few moments later the natives again emerged, bringing with them two tiny men, who struggled fiercely to escape their captors. They were pygmies—the first seen at close quarters by a white man. They were given presents of colored beads and later released, as Rawling wished to keep on friendly terms with these jungle men. It was fortunate that he did so, for he later ran into an ambush, and his life was spared only by the intervention of the pygmy whom he had formerly released.

Progress through the jungle was exasperatingly slow. Every foot of the way had to be hacked out, clearing the dense underbrush. Poisonous snakes and insect pests were everywhere, and many of the porters died of malaria and beri-beri. Great difficulties were experienced in bringing provisions up from the base camp. River traffic was next to impossible, owing to rapids, whirlpools, and floating logs. Visibility was limited to a few yards ahead, and the direction, in the darkened jungle, had to depend almost entirely upon the compass. Despite these obstructions and difficulties, slow progress was made, and the party finally came to another river, the Iwaka.

Unfortunately this proved to be a raging torrent, and any passage across it seemed impossible. Yet it had to be crossed if the journey was to be continued. Finally one of the Gurkhas undertook to swim across, with a rattan rope tied around his waist. This he fastened to a tree on the other side. Then another Gurkha swam across, clinging to the frail life-line. The current was so swift that his body was level with the surface of the water most of the time, and he ferried himself across by means of his hands. Finally, however, he reached the opposite bank, and then additional ropes could be stretched across, rendering the passage of men and supplies possible.

Then the slow progress was continued—hacking their way through the dense jungle, struggling against appalling difficulties. To the north, Mount Godman could be seen, from which a good view of the surrounding country could be obtained, so that maps of it could be made. Finally, after days of unremitting toil, the summit was reached; but it was then found that a heavy mist rendered distant visibility impossible. There was nothing to do but wait until the mist lifted. A few mornings later a clear view of the country suddenly came into view. At their feet lay the jungle from which they had just emerged, and to the north a great precipice stretching for 80 miles, with an unclimbable scarp nearly two miles high. Quickly rough maps were drawn, and a clear idea of the country obtained. But it had been at an enormous cost of human life. Out of a total of 400 coolies, convicts and other servants who had been taken to the island, only 11 remained. All the rest had died of fever, beri-beri or exhaustion. However a huge area had now been mapped, which rendered subsequent surveying of the island possible.

Similar hardships and difficulties were experienced by other explorers of New Guinea—Staniforth Smith, who first traveled into the interior of Papua (the British portion of New Guinea): Michael Leahy who, during the years 1930-34, made 10 journeys into the central plateaux, where he encountered the cannibals of the highlands—and several others. These initial explorations among the head-hunters frequently involved grave dangers, and on a number of occasions attacks by hostile natives had to be beaten off with considerable loss of life. In later years, airplanes were effectually used for mapping the interior.

EXPLORERS OF THE AMAZON—RIVER OF DOUBT

The Amazon River has always held a certain fascination for explorers, and public attention was attracted to it years ago by Theodore Roosevelt's journey up the River of Doubt, and by the dramatic disappearance of Colonel Fawcett, whose fate still remains a mystery. The latter had had much experience in South America, and had been engaged to determine the boundaries of Bolivia, Peru and Brazil, which he did in the early years of the present century. Later he made a number of trips into the interior—his last fatal journey being undertaken in 1925, accompanied by his son Jack and an Englishman named Raleigh Rimell. In a message sent back on May 20th, he said:

"From Fort Bakairi, whence I sent my previous dispatches, our journey has been no bed of roses. We have cut our way through miles of *cerraba*—a forest of low, dry scrub; we have crossed innumerable small streams by swimming and fording; we have climbed rocky hills of forbidding aspect; we have been eaten by bugs . . . Our two guides go back from here. They are more and more nervous as we push further into the Indian country."

A year passed, then two. No word was received from Fawcett. Finally, in 1928, a relief expedition was organized by Commander Dyott, who had previously explored much of the interior of South America. He and his party pushed their way up winding rivers, infested with huge watersnakes, and across dangerous country, filled with hostile Indians. Old camp sites were found and definite relics of the Fawcett party, but it was impossible to find his grave, or any definite clue as to his ultimate fate. Rumors continued to be circulated for some years that he was still alive, but none of these could be verified. It is now practically certain that he and his entire party were killed by the Indians somewhere in the interior of mysterious Matto Grosso.

EXPEDITIONS TO CENTRAL ASIA AND TIBET

Far on the other side of the world lie the great wastes of central Asia and Tibet, virtually inaccessible to white men—until Colonel Younghusband's military mission to Lhasa in 1903. Even today it is one of the most forbidding and difficult countries in the world to enter or explore. A few individuals had, from time to time, penetrated its mysteries—Sven Hedin being the most famous of these. His journey into the interior was made as far back as 1893, and thenceforward, for many years, he led expedition after expedition into Tibet. His amazing, almost incredible adventures may be found in his books—for which reason they will be touched upon only briefly here.

His first (1893) journey was a tragedy. Approaching Tibet through the great Tekla-Makan desert, composed of mountainous sand dunes, the water supply completely gave out. The camels died one by one, and the men dragged themselves forward a step at a time, until they too fell exhausted. They drank the camel's rancid oil, and Hedin himself drank some Chinese brandy, which would normally have been used for a lampstove. This paralyzed his muscles and he dropped far behind the rest, crawling forward on all fours. The men died until only Hedin and his faithful servant, Kasim, were left. The latter was delirious. Mile after mile they crawled forward, until at last they came to the wooded banks of a desert stream. But they were too weak to reach it. Lying in the shade of a poplar, they waited for the cool of the evening before making the attempt. By that time Kasim was unable to move, but remained where he was, lying on his back, motionless, with his eyes wide open, unable to speak. Hedin, using his spade-handle as a staff, crawled forward until he came to the river bed—only to find it dry. As despair seized him, a duck flew into the air and water splashed. Evidently there must be a pool somewhere. Crawling forward again he finally discovered it—fresh, clear water, the one pool that was left. He drank deeply, then filled his top-boots with water, which he took back to Kasim. Soon afterwards they were both rescued by some passing shepherds.

In later expeditions, Hedin discovered the ruins of many buried cities. In one of these, Loulan, they found papers inscribed with early Chinese ideographs, or picture-writing. When these were interpreted, they were found to give a clear picture of life in Loulan in the year A.D. 270. It had been a flourishing frontier town on the great Silk Road between China and the Roman Empire. The discovery of some mollusc-shells not far away showed that it must once have stood near the shores of a lake.

In 1905, Sven Hedin actually reached Shigatse, where resided the Tashi-lama, second only to the Dalai-lama of Lhasa in importance. Here for some strange reason he was allowed to remain for 47 days, moving freely about the town, taking photographs and witnessing religious ceremonies. Among other oddities he visited a kitchen where, in six enormous caldrons, tea was brewed for 3,800 monks. Eventually, however, a message came from Lhasa, intimating that he had overstayed his welcome, and he was compelled to leave. During the course of his various expeditions, Hedin had explored many hundreds of miles of totally unknown territory, and added much to the world's knowledge of present-day Tibet.

AUREL STEIN IN CHINESE TURKESTAN AND CAPTAIN RAWLING ON THE GREAT PLATEAU

Less well known are the explorations of Aurel Stein, in Chinese Turkestan, and of Captain Rawling on the Great Plateau. Yet their adventures too are of thrilling interest. Stein's most amazing discovery was at Hun-Huang, where he was permitted to pay a visit to the Caves of the Thousand Buddhas. In this sacred valley he found innumerable grottos, most of them empty save for images of the Enlightened One. The walls were covered with paintings, images and frescoes. In one of these caves, guarded by an old Taoist priest, was an anti-room, and on looking into it Stein was amazed to see a stack of ancient manuscripts almost 10 feet high.

At first the old priest was reluctant to show any of the texts, evidently in fear of being betrayed by fanatical pilgrims. But at length Stein was allowed to examine several of them, and he then intimated that he would be willing to make a liberal donation to the grotto if he might be permitted to take some of them away for further study. Finally a bargain was struck, and for four silver horseshoes Stein was allowed to remove and ship home 29 cases of manuscripts and art treasures—which were sent to the British Museum. Stein added greatly to these treasures later, and on his return to London brought with him enough material to keep the archeologists of Europe busy for many years.

Rawling's crossing of the Great Plateau must be passed over with slight mention, for, though it was filled with dangers and innumerable hardships, and though he was repeatedly turned back from his objective, Rudok, by angry nomads, when almost there, still his experiences were similar to those of many other explorers in the vast interior of Tibet. The same may be said of his Gartok expedition which, though full of interest, must be omitted here.

THOMAS AND PHILBY JOURNEY ACROSS THE ARABIAN DESERT

One further series of desert explorations should, however, be mentioned. The southern portion of Arabia consists of an enormous waterless desert, known to the natives as the Abode of Death. This is Rub al Khali, and is as large as France and Spain put together. Until the beginning of the present century this vast area had never been crossed by Europeans; then, within a single year, it was crossed twice by explorers. Bertram Thomas and H. St. John Philby both negotiated the journey successfully, traversing the desert from north to south and from east to west. It was a remarkable coincidence, since neither of them knew of the intentions of the other.

Thomas knew the settled parts of the country well, and spoke the local dialects perfectly. In fact, he had been appointed Wazir to the Sultan of Oman, and his name was everywhere respected. Nevertheless he had to keep his project to himself, as it would have won official disapproval. In December, 1930, Thomas set forth, guided by a Bedouin chieftain, Shaikh Salih. There was war in the desert between the Rashidis and the Sa'ar, which rendered his journey all the more perilous. The first 100 miles lay along the southern fringe of the sands, where there were a few oases—but also marauding bands of Bedouins. Then the party turned into the desert proper.

Progress was slow, but they at length reached a place called Dhahiya, where they met a Murra tribesman who agreed to guide them through the northern part of Rub al Khali. Their final dash across the desert was to take them to Doha, on the Persian Gulf, some 350 miles distant. Hunger, thirst and sand storms plagued them—as they have all those who cross the desert. Had it not been for these, Thomas might have made an astonishing discovery. One day the Arabs pointed to some deep tracks in the sand, saying that these led to the city of Ubar—now buried beneath the sands. Was Ubar but the Arab name for Ophir, that fabled city "from which were fetched 420 talents of gold, which were brought to King Solomon"? (I. Kings, 9). The Arabs assured Thomas that they had visited the sand-covered city, and had found pottery and seen the tops of columns projecting above the earth. Reluctantly however Thomas had to abandon the idea of exploring the site of the ancient city at that time, due to shortage of water.

Thomas completed his journey a few days later, arriving safely at Doha, after having made the first trip across the famous Arabian Abode of Death.

Philby, who also knew Arabia well, was fortunate enough to travel under the patronage of the King of Hejaz. His caravan, consisting of 15 specially bred camels, covered 1,800 miles of blistering desert in 90 days— a remarkable record considering that the beasts had to be driven mercilessly and were at one time 10 days without water. Moreover, it was bitterly cold at night, and every morning the water-skins had to be thawed out, before the ice could be converted into drinking water.

Philby began his journey at the northern edge of the desert, traveling southward. His guides also told him the legend of the buried city of Ubar. Philby decided to explore the spot himself, his interest having been further aroused by the discovery of some fresh water shells and ancient flint implements in the sands. Arriving shortly before dark, he hastened to the top of a sand-hill and looked down on the ruins of what appeared to be a volcano. "Below me," he wrote, "as I stood on that hilltop transfixed, lay the twin craters, whose black walls stood up gauntly above the encroaching sand like the battlements and bastions of some great castle."

This discovery, however, almost cost Philby his life.' Their supply of water ran out, and the thirsty camels had to be driven for almost 24 hours continuously before the first signs of vegetation were noted. This last forced march broke the back of the journey, however, and by evening they had reached water and the first outposts of civilization. The desert of Rub al Khali had been successfully crossed for the second time within 12 short months.

EXPEDITIONS TO MT. EVEREST, THE "ABODE OF SNOWS"

Of all the explorations which have ever been undertaken, however, none has seemingly fascinated man's imagination more than the ascent of Mount Everest, the highest peak in the world. Attempt after attempt has been made to reach the summit, only to end in disaster and frustration. One can perhaps understand the belief of the Tibetan lamas—that the gods of the snows guard it from all intrusion—and because of this refused all requests by explorers to attempt its ascent. In 1920, however, the Royal Geographical Society obtained permission to enter Tibet and explore the approaches to the mountain. Three attempts to climb it were made within three years, and then the Tibetan authorities refused permission for eight years. From 1932 onwards, however, a number of additional attempts have been made, and planes have flown over it a number of times, taking photographs of the mountain top, the approaches and the surrounding country. These proved of great value in subsequent expeditions.

Mount Everest is slightly more than 29,000 feet in height, and lies at the southern edge of Tibet, just north of Nepal. It is surrounded by huge glaciers, which extend for miles, and the sides of the mountain are often sheer rock, thousands of feet in height, which do not even afford a hand-hold. The southern face is quite unclimbable, not only because of these features, but also because it is exposed to the full fury of the monsoons, which approach from this direction. The northern slopes are slightly more gradual, but these are ice-coated and covered with soft, treacherous snow. To the northeast is a ridge of ice known as the North Col. This is seamed with crevasses and ice cliffs. In addition, the strata of the mountain often point downward, like the tiles on a roof, rendering frequent detours necessary. The ledges are narrow and slippery, and hand-holds have to be cut with an ice-axe. Winds are treacherous, unexpected snow falls and avalanches frequent, and one false step may send the explorer crashing into a ravine thousands of feet below. In addition to all this, the intensely rarefied air makes the slightest exertion difficult and the heart pound like a trip-hammer. Zero temperatures are frequent and snow blindness is an ever-present danger. It is small wonder, then, that none but the most hardy and experienced mountaineers have attempted the ascent—and then only after a careful period of acclimatization.

In making the attack upon Mount Everest a base camp is first established; then a series of camps numbered respectively 1, 2, 3, etc., as the gradual ascent continues. These camps are as essential in mountain climbing as they are in polar exploration.

The first expedition, led by General Bruce, reached a point of almost 27,000 feet; then the monsoon broke and they had to beat a hasty retreat. Two years later, Bruce led a second expedition, which included a number of experienced climbers—Mallory, Norton, Odell, Finch and Somervell. The leader, Bruce, was stricken with fever, and one of the others with dysentery. Finally, however, Camp 5 was established at 25,200 feet, despite a terrific hurricane. Beyond this point the porters refused to go.

After an argument which lasted for four hours, they were finally persuaded to go on, and Camp 6 was established at 26,800 feet. Every step required fearful exertion, and some of the porters had to be returned to the base camp.

On June 4, Norton and Somervell began their assault on the summit. At 28,000 feet the latter collapsed. Norton, though suffering from snow blindness and seeing double most of the time, with great pluck pushed on, only to be stopped 100 feet higher by projecting rocky slabs. Luckily both he and Somervell reached the camp in safety.

A day or so later, Mallory and Irvine made another attempt—after sending back a note to Odell, at Camp 5, that they were about to do so. What happened no one will ever know. Odell set out in support, and as the mist lifted he seemingly saw, for a brief moment, a dark figure on the very last step, and another a little below it. Then the mist again descended. Nothing more was ever seen of the two intrepid adventurers. Odell made a heroic effort to reach them and, though alone and numbed with cold, ascended to more than 27,000 feet before being compelled to turn back. Were the two black specks which Odell saw an illusion, or did one or both explorers actually reach the summit of Mount Everest—only to disappear forever? Here indeed is a mystery—which has remained unsolved until the present day.

In 1932-3, a fourth Everest expedition was organized by Hugh Ruttledge, including a team of wonderful climbers: Smythe, Shipton, Harris, Birnie, Brocklebank, and others. First the wall of ice leading to the North Col was climbed. Then six camps were prepared and provisioned, the last one at a height of 27,400 feet—perched on the edge of a three-foot ledge, so that part of the tent projected over it into space. Moreover the floor sloped dangerously. However there was no other available site, and Camp 6 was established there.

Harris and Wager remained, while Lomgland led the tired porters back to Camp 4, through a blinding blizzard. Early the next morning the two explorers began their great effort to reach the summit. They rounded the first Step and reached the Second, only to find that the hard, gray rock afforded no hand-holds. They continued to circle the ridge crest, finally coming upon 50 feet of snow, which had to be crossed before the ascent could be continued. Cautiously they crept forward; a snow-slide would have carried them down to the Rongbuk glacier, 10,000 feet below. From this point the ascent looked possible; but it was now afternoon, and they had to return to camp before dark, so reluctantly they turned back.

On reaching Camp 6, they found Smythe and Shipton, who were to form the second assault party, already there; so, passing on what information they had acquired, they descended to a lower camp. For two days Camp 6 was completely snowbound, but on the third day the weather cleared, and Smythe and Shipton began the ascent. Soon the latter complained of stomach trouble, and was unable to go further. This was a bitter blow, since it was almost certain that Smythe could not reach the summit alone. However, he pushed on and had gained 50 feet during the course of the next hour. Then he found that newly fallen snow rendered further progress impossible, and he was forced to return. He reached Camp 6 by dusk, to find Shipton alive and well.

At Camp 4 another assault was now planned, but snow fell heavily for two days, and it soon became evident that the monsoon had broken. This meant that further attempts were impossible in the immediate future; so on June 21 the camps were abandoned and the homeward march begun. Once again the ice-demons had driven the white men from the Abode of Snows.

In 1936, another expedition was organized, also under Hugh Ruttledge, but mishaps, bad weather and accidents dogged their attempts from the beginning, and the explorers never succeeded in reaching the heights attained by former climbers. Then came the Great War. And so the summit of Mount Everest still remains unattained, its enigma as yet unsolved. The bodies of many brave men lie somewhere hidden beneath its snows, but the Tibetan gods of the mountain still remain undisturbed, as they have these many centuries.

Well might we say, with Coleridge, in his Kubla Khan:

A savage place! as holy and enchanted
As e'er beneath a waning moon was haunted
By woman wailing for her demon-lover.

TWO FAMOUS WOMEN EXPLORERS

ROSITA FORBES AND "THE SECRET OF THE SAHARA"

.Women have frequently accompanied their husbands in dangerous explorations, but few of them have ventured alone into unchartered wildernesses. The reasons for this are obvious. Their sex in itself exposes them to dangers from which a man would automatically be exempt, ranging all the way from assault to kidnapping. Then, too, such explorations usually involve tremendous endurance, physical strength and stamina, no less than pluck and determination. Under the circumstances, it is only natural that the majority of our explorers should have been men. All the more credit, then, to those extraordinary women who underwent the same dangers and hardships. It is strange indeed that their names are known to so relatively few. I propose to recount, briefly, the discoveries and adventures of two such women.

Few realize the dramatic immensity of some of the relatively unexplored regions of the earth. The Amazon River, for example, carries between its jungle-clad banks one-10th of all the river water of the world. Two thousand miles from its mouth the bed is still deep enough to permit ocean-going steamers, and its gigantic basin is an area almost the size of Europe. Again, Africa is an immense continent, in which the United States, Canada and the whole of Europe might be tucked, with millions of square miles to spare. The Sahara Desert alone covers an area larger than the continent of Australia. Until our own generation, this immense region was almost as inaccessible to white men as Tibet.

This vast area (the Sahara) is for the most part governed by the *Senussi*—a collection of wild tribes, followers of a religious fanatic born just 160 years ago. Sidi Mohammed Ben Ali Es Senussi was a mystic—a sort of Mohammedan St. Francis. His followers now run into the millions.

Kufara, their sacred city, has a reputation second only to that of Mecca. It has always been closely guarded, and is known as "The Secret of the Sahara." In 1879, a German scientist named Rohlfs attempted to reach the city, but failed and barely escaped with his life, as the Senussi had a habit of shooting all strangers at sight—particularly the hated *Nasrani* (Christians). For more than 40 years after the Rohlfs expedition, Kufara was left strictly alone, while the rest of the world went its merry way.

Then, in 1920, an English lady who had traveled extensively in many wild regions of the world, made up her mind to penetrate "The Secret of the Sahara." She would succeed where mere man had failed. To be sure, she took a companion explorer with her—a man who spoke the language, understood the people and knew something of the country. But she was the leader of the expedition nonetheless.

This traveling companion of hers, Hassanein Bey, was a strange mixture of the civilized and the barbaric. He was the son of an Egyptian noble and had graduated from Oxford University, where he had won his "blue." While living like any Arab nomad, he nevertheless considered bath-salts, Eau-de-Cologne, blazers and patent-leather shoes "necessities," and never traveled without them. Yet in many ways he was a true Son of the Desert.

At Benghasi—immortalized in World War II—Mrs. Forbes had her first stroke of luck, for here she was fortunate in meeting Idris Es Senussi, the ruling Emir, who wrote her a cordial letter of welcome, which later acted as a magic passport whenever shown in the interior. Of course, she did not explain her objective to that notable, but stated merely that she wished to explore the deserts—in this way doubtless avoiding his direct veto. Kufara lies nearly 600 miles south of Benghasi, and days of waterless waste lie between. Only the best-equipped native caravans attempt the trip, and then only on rare occasions.

Despite these seemingly insurmountable difficulties, Rosita Forbes set forth, and in a few days reached Jedabia, about 100 miles south of Benghasi, where she was welcomed by the Emir's brother, and for some days was lavishly entertained. "They were given a house, a cook, and servants, and invited to a feast of soup, chicken, mutton, tomatoes, marrow, rice, omelettes, and mint-tea, which lasted for three hours."

They soon found, however, that the town was infested with Italian spies, and plans had to be discussed with the utmost secrecy. Difficulties were encountered in obtaining camels, for their vague "trip into the desert." Moreover, rumors began to be circulated that she was a wealthy Christian lady, ripe for plunder by marauding bands. Rosita Forbes attempted to offset this by reciting Moslem prayers and verses from the Koran in public. She had already adopted the traditional Bedouin dress —tight white trousers, a red gown, with a strip of woolen material wrapped 'round the hips. She was careful, however, to conceal beneath this two loaded revolvers and a prismatic compass.

Nevertheless, suspicions began to be more and more audibly expressed. Why was this white foreigner remaining so long in a small mud village far from civilization? Questions began to be asked, which they found it increasingly difficult to answer. Finally it was decided that the safest course was flight, and this was arranged for the following night. The Emir's brother instructed two of his servants, Yusof and Mohammed, to act as guides and accompany them "to the end of their destination." It was fortunate indeed that he did so, as subsequent events proved.

Late that night, in inky blackness, the camels were furtively loaded and the journey south was begun. At first, Yusof lost his way, and when dawn broke they found that they were still within sight of Jedabia. Hastily the flight was resumed—the camels being urged on to their topmost speed. Despite a strained ankle, which caused her intense suffering as the camel swayed about under her, the pace was never relaxed until it was felt they were safely launched on their journey.

Two days later they reached the *wadi* Farig—a few scattered tents around a desert well. Here they met a merchant by the name of She-ib, and, finding that his caravan was going southward, they joined forces with him and continued their journey—rather than wait for a promised caravan, which had been sent to support them. Together they reached the next oasis, Aujela, after battling terrific sand storms, which blinded the camels, causing them to swing in circles. The sand penetrated the food boxes, covering everything with a fine grit. The water-skins were empty by the time they reached Aujela.

Here, however, they found the promised caravan awaiting them—12 camels and a dozen men under a famous guide, Abdullah. Friction soon developed between the black soldiers and the Arabs, and tragedies were only prevented by sheer diplomacy. The lives of explorers have only been saved by the quick exercise of wits, times without number.

The next oasis, Jelo, was reached without mishap, but from that point on their troubles began. The next stage of the journey involved a trek over 250 miles of waterless desert. Further, it soon became evident that Abdullah, despite his reputation, had lost his way. The food was almost gone and only one day's supply of water was left. "The camels were famished and tried to eat the stuffing of the baggage sadles; it was their ninth day without water."

In this crisis Rosita Forbes decided to ignore the advice of Abdullah, and strike out southeast toward Buseima. The bedouins were hesitant, as they had heard grim tales of this district. Nevertheless there seemed no alternative, and the tired camels were once more driven on in this new direction. That evening they came upon some green bushes, and their hopes rose high—only to be dashed by Abdullah, who informed them that this spot was known as "The Thirst," and that the water there was poisonous. So the last dry supper was eaten, the camels were given the stuffing from the sadles, and the parched travelers tried to sleep, hoping that the next day would bring them—water.

Fortune favors the brave. The following afternoon a few scattered palms were sighted, and the soldiers had soon dug a hole, which slowly filled with muddy water. On this man and beast quenched their thirst. Then the journey was resumed. Two days later Rosita Forbes and her caravan reached Buseima.

Owing to the ominous forebodings of the natives, trouble had been expected there, but when they reached this oasis the party was well received, owing perhaps to the small number of the inhabitants, which did not equal that of the caravan. However, this apparent friendliness was only a cloak for inward suspicions and resentment. As a matter of fact, immediately upon their arrival, a messenger had been sent to Kufara, who informed the Governor that the Emir had deceived his own people, had sold Kufara to the Italians, and that the travelers were only an advance guard of an invasion to follow. (As it turned out, this proved to be quite prophetic, for 10 years later the Italians occupied Kufara.)

Word was accordingly sent back that their entry into the holy city would not be permitted. Yusof, to whom this message was given, felt that some traitor had betrayed and misrepresented them, and set off post-haste to Kufara himself. He explained the true situation, and later returned, triumphantly waving a piece of paper. This proved to be a letter of welcome from the Governor, and permission to enter the holy city. Excitement ran high as Rosita Forbes and her caravan set off on the last leg of their eventful journey.

The following day they reached the crest of a low, black hill, and there, below them, they saw a beautiful, hill-encircled *wadi*. Scattered villages were seen among the mile-upon-mile of palm trees, while at Taj —the principal village—towering above the small, dark houses below, was seen the Governor's Headquarters. Green gardens surrounded a beautiful blue lake, and the whole scene reminded them of a lovely jewel, set in the midst of the desert. The beauty of the scene was breathtaking.

An so, at long last, the object of her ambition was realized, as Rosita Forbes with beating heart rode at the head of her caravan into Kufara, "The Secret of the Sahara," which before no white man had ever seen. Assuredly it was a proud day for her. Almost awe-struck, they gazed about them at the quaint, picturesque beauty of the villages, and the surrounding fields, where fig-trees, palms, olives and vines grew in profusion. The letter from the Emir stood her in good stead, for she was cordially received, and permitted to visit all six of the native villages, and was also shown the interior of the sacred places and the Governor's palace, which consisted of many courts and intricate passages. Paradoxically enough, these were partially furnished with articles of European make, and in one room Mrs. Forbes counted no fewer than 15 clocks.

Some days were spent in Kufara before the hazardous return journey was begun. Again this had to be in the nature of a semi-flight, for though the officials were friendly the natives were not. But it was safely begun and, after further waterless days and nights, and a series of minor adventures, the caravan finally reached the outposts of civilization and disbanded. Rosita Forbes accomplished the seemingly impossible—a visit to the sacred city of Kufara, which no foreigner had ever before gazed upon.

FREYA STARK'S EXPLORATIONS IN PERSIA

If the explorations of Rosita Forbes are dramatic, those of Miss Freya Stark are even more so, since they were undertaken almost single-handed, with only local guides and retainers, whom she picked up en route. A student of Oriental customs and languages, and an archeologist of some repute, Miss Stark's journeyings were for the most part through northern Persia, consisting of vast wastelands of unexplored territory, infested with bands of robbers, and considered so dangerous that even the police and government officials hesitated to visit it. Freya Stark had explored many of the wild regions of Persia before, but in 1930-32 she undertook an expedition which for daring has but few equals in the history of exploration.

Setting out from Bagdad and heading toward Luristan, she encountered a young Lur who told her an extraordinary story of a vast treasure, hidden in a cave in the hills of Kebir Kuh (south of the Caspian). He supplied her with a rough map and was to accompany her, but at the last minute his passport was confiscated by the authorities and he was prevented from doing so. Nothing daunted, Freya Stark set off by herself, without the young Lur. Her only guide was an old quiltmaker who knew the surrounding country.

When they reached Kebir Kuh, her guide told her of the ruins of two ancient cities—Larti and Hindimini. So, turning aside from what she called her "Arabian Nights Adventure," Miss Stark decided first of all to explore them. The ruins of both cities were still plainly visible, the streets and some of the walls rising above the dense underbrush. Many of the tombstones, covered with early Persian script, were also recognizable. No treasure was found, but Miss Stark carried away with her a human skull, carefully wrapped in her Burberry.

At this juncture, her plans to discover the hidden treasure cave were rudely interrupted by the sudden appearance of three mounted policemen, who had been sent to find her—word having been received by the authorities that a strange white woman was in the vicinity. To tell them the object of her search was out of the question. They would have forbidden it, since the intervening hills and ravines were infested with brigands. But Freya Stark was not to be deflected from her purpose by any such minor difficulties. She put her wits to work. She prepared a full lunch, provided cigarettes, and then suggested that the men take a brief nap while she went off to look at some nearby ruins. Her stratagem succeeded. While the men were asleep, Miss Stark headed for the ravine in which (according to her map) the cave was situated, and for two hours clambered over the slippery rocks, looking for the entrance to the cave. She could not find it and, knowing that her escort would now be looking for her, she reluctantly gave up the search and returned to the camp. If such a cave really exists it is still to be found—though a tribesman later informed her that it had long ago been rifled and that there was now nothing inside it.

Freya Stark's most romantic journey, however, was her visit to the "Valley of the Assassins," and the retreat of the "Old Man of the Mountains." This was situated in the mighty Elburz range—only a few hundred miles from the scene of her former explorations.

The legend of the Old Man of the Mountains may be found throughout medieval history; the Crusaders knew it, and throughout the middle ages the mere mention of it struck terror in the minds of Moslems and

Christians alike. The founder of this fanatical religious sect, according to tradition, lived in the fastnesses of some inaccessible region, from which raids were made by his folowers into the surrounding country. The Old Man was said to possess a secret garden where a mysterious herb was grown, which induced ecstatic visions in those who partook of it. It is now known that this was hasheesh, and the sect accordingly came to be known as *Hashishin*—from which our name "Assassin" is derived.

The palace in which the Old Man lived was said to be a mighty castle, perched on nothing less than Solomon's Throne. Though none had ever seen it, the Valley of the Assassins was, of course, known to the authorities. Freya Stark determined to find the lair of the Old Man of the Mountains, and beard his living representative in his den.

It was a perilous adventure, which had to be undertaken with the utmost secrecy. Starting her journey from Qazvin, she eventually reached Chala Pass (8,000 feet) beyond which was a long ravine. Emerging from this she came to the valley of Alamut, and saw in the distance a vast rock, on which stood the castle of the Old Man. She found that the valley narrowed into a ravine, with a rock wall on one side 3,000 feet high. At the end of the ravine was a high pass, leading to dense forests running all the way to the Caspian Sea. Evidently the Old Man had built his castle surrounded by almost impregnable defenses.

Freya Stark was conducted up the steep rock to the ruined castle by a red-bearded old Assassin. It was now empty and desolate, the rocks and ancient walls, built on the edge of a sheer precipice, being overgrown with wild tulips. She found to her surprise that the Assassins, though still existent, are now a peaceful sect, ruled over by their official head, Aga Khan. Raids were a thing of the past, and the hillmen, though wild, were no more so than those of the surrounding country. She found little in the present ruins suggestive of the castle's grim history, or the terror which it inspired when the Old Man of the Mountains was still alive. Like all cruel and tyrannical organizations it had withered and perished in the inevitable passage of time.

The following year (1933) Freya Stark was awarded the Back Grant by the Royal Geographical Society for her discoveries in Persia. She had traversed and made maps of regions where even the Persian officials were afraid to go. She shares, with Rosita Forbes, the distinction of being numbered among the world's great explorers—both having visited for the first time localities which no white man had ever before reached or penetrated.

H-J BOOKS BY DR. HEREWARD CARRINGTON

Dr. Hereward Carrington was born in St. Herlier's, Jersey, Channel Islands, on October 17, 1880. His father, Robert Charles Carrington was connected with the British admiralty all his life. His mother (Jane Pewtress) was Polish; his father's ancestry Irish; both British. Educated in England, Dr. Carrington came to the U.S. in 1899 and remained ever since. He became a naturalized citizen. He has lived mostly in New York City; now in Hollywood, Calif. His first job was with Brentano's book store. For a time he was editor of Street and Smith's 10-cent novels. Later he became a free-lance writer and lecturer. He has written (big and little) over 100 books and more than 1,500 articles and reports. He has been an amateur magician all his life. Hobbies: tennis, science, bridge. In addition to writing motion pictures and playlets, Dr. Carrington has been on the radio for many months. He has traveled extensively through Europe. His list of books for Haldeman-Julius follows:

LITTLE BLUE BOOKS BY DR. HEREWARD CARRINGTON

(10c each; delivered anywhere in the world.)

419	Life: Its Origin and Nature	409	Great Men of Science
524	Death: and its Problems	1321	Fasting for Health
417	The Nature of Dreams	761	Food and Diet
491	Psychology for Beginners	1277	Hindu Magic Self Taught
895	Astronomy for Beginners	1285	Gamblers' Crooked Tricks
679	Chemistry for Beginners	1279	Side-Show Tricks Explained
493	Novel Discoveries in Science	1278	Ventriloquism Self Taught
602	The Great Pyramids of Egypt	421	Yoga Philosophy Explained

LARGER BOOKS BY DR. HEREWARD CARRINGTON

Perfumes—Their Sensual Lure and Charm. 35c.

The Psychology of Genius. Why some have faculty for original, brilliant and creative work. 25c.

Fears—And How to Banish Them. Practical help for sufferers from fear neuroses. 25c.

Valuable Health Hints. Suggestions for living a sane, normal, wholesome life. 35c.

Your Eyesight. An outline of the Bates Method of treatment without glasses. 25c.

The Seven Wonders of the World. Ancient, Medieval and Modern. 25c.

More Scientific Oddities. Numerous helps to general information. 35c.

Scientific Oddities. Little-known facts, paradoxes and illusions, puzzles and quizzes, etc. 25c.

The Book of Rogues and Impostors. Historical and critical summary of legends, swindles, hoaxes & rackets. 25c.

The French Menu. How to read and understand it. 25c.

How to Live. Helpful thoughts on a sound philosophy of life. 35c.

Little Known Explorations. Overlooked or forgotten discoveries that nevertheless have furnished facts that have served to unlock the secrets of the past. 25c.

Psychology of Salesmanship. Practical guide to successful selling. Carrington. 25c.

All 16 Little Blue Books and 13 Larger Books may be had at the special price of $4.70, prepaid. Ask for: 29 BOOKS BY DR. HEREWARD CARRINGTON. Mail orders to:

HALDEMAN-JULIUS PUBLICATIONS, GIRARD, KANSAS

Books by Upton Sinclair

The following books by Upton Sinclair, bearing the Haldeman-Julius imprint, are available for immediate delivery to all customers who mail us their orders immediately:

1. The Book of the Mind, 75c. 2. The Book of the Body, 75c. 3. The Book of Love, 75c. 4. The Book of Society, 75c. These four volumes constitute Upton Sinclair's "The Book of Life." All four volumes, $2.

American Outpost. A book of reminiscences. 82,000-word autobiography. $1.50.

Limbo On The Loose. A midsummer night's dream. A new story that looks at what lies ahead for America, and the way out. 60c.

A Giant's Strength. Dramatic story of atomic bomb, its past, its present, and one among its possible futures. Humorous, sophisticated, witty, charming story, with screamingly funny jabs at radio commercials. 60c.

Boston—800-page novel of the Sacco-Vanzetti case. $2.

Oil!—A novel. 525 pages. $2.

The Goose-Step—A study of U. S. Education. 500 pages. $2.

The Brass Check—A Study and Exposure of American Journalism. 446 pages, 2 vols. $1.50.

Profits of Religion. Supernaturalism as a source of income and a shield of privilege. $2.

Is the American Form of Capitalism Essential to the American Form of Democracy? Debate between Upton Sinclair and George Sokolsky. 25c.

No Pasaran! (They Shall Not Pass). A novel of the battle of Madrid. 50c.

Letters to Judd—An American Workingman. 50c.

The Flivver King. A novel of Ford America. 60,000 words. 50c.

Peace or War for the U. S. A. Debate between Upton Sinclair and Phil LaFollette. 25c.

What Can Be Done about America's Economic Troubles? 25c.

Expect No Peace. 25c.

Your Million Dollars. 25c.

The Cry for Justice. Anthology (abstracts) of social Protest. 25c.

Can Socialism Work? 10c.

The Jungle. A novel of the Chicago stockyards. 60c.

The Pot Boiler. 10.

The Millennium. 30c.

The Second-Story Man. 10c.

The Naturewoman. 10c.

The Machine. 10c.

Captain of Industry. 20c.

Socialism and Culture. 10c.

Also: A book about Upton Sinclair, by Joseph McCabe, entitled "Upton Sinclair Finds God." 25c.

If you want all 31 books listed above remit $14.45 (a saving of $6) and ask for COMPLETE LIST OF TITLES BY UPTON SINCLAIR AS PUBLISHED BY US. If you order less than complete set, remit as priced above after each title. All Sinclair books, whether ordered in complete sets or selections of titles, are shipped carriage charges prepaid. Mail your order and remittance to:

HALDEMAN-JULIUS PUBLICATIONS, GIRARD, KANSAS

CPSIA information can be obtained
at www.ICGtesting.com
Printed in the USA
BVOW06s2240100917
494514BV00019B/287/P